ENVIRONMENTAL REMEDIATION TECHNOLOGIES, REGULATIONS AND SAFETY

PHYTOREMEDIATION TECHNOLOGIES FOR THE REMOVAL OF TEXTILE DYES: AN OVERVIEW AND FUTURE PROSPECT

ENVIRONMENTAL REMEDIATION TECHNOLOGIES, REGULATIONS AND SAFETY

Additional books in this series can be found on Nova's website under the Series tab.

Additional E-books in this series can be found on Nova's website under the E-book tab.

CHEMISTRY RESEARCH AND APPLICATIONS

Additional books in this series can be found on Nova's website under the Series tab.

Additional E-books in this series can be found on Nova's website under the E-book tab.

ENVIRONMENTAL REMEDIATION TECHNOLOGIES, REGULATIONS AND SAFETY

PHYTOREMEDIATION TECHNOLOGIES FOR THE REMOVAL OF TEXTILE DYES: AN OVERVIEW AND FUTURE PROSPECT

SANJAY P. GOVINDWAR
AND
ANURADHA N. KAGALKAR

Nova Science Publishers, Inc.
New York

For permission to use material from this book please contact us:
Telephone 631-231-7269; Fax 631-231-8175
Web Site: http://www.novapublishers.com

LIBRARY OF CONGRESS CATALOGING-IN-PUBLICATION DATA

Available upon Request
ISBN: 978-61761-746-1

Published by Nova Science Publishers, Inc. † New York

CONTENTS

PREFACE

Phytoremediation which involves the use of plants and rhizospheric organisms for the removal of pollutants is an emerging technology for the clean up of contaminated sites. The removal of textile dyes mediated by plants has been one of the most neglected areas of phytoremediation research. Dyes, which are primary constituents of the wastes from textile industry effluents, constitute a group of recalcitrant compounds, many of which are known to have toxic and carcinogenic effects. Hence, the review focuses on the studies of the mechanisms adopted by plants in the removal of textile dyes and the future scope for research in this area which will help in broadening the horizons of phytoremediation technologies. Plant species many a times referred to as 'green livers', are known to possess a wide range of detoxifying and biotransforming enzymes some of which may also be secreted extracellularly in the rhizosphere and can bring about the transformation of organic pollutants such as textile dyes. The use of *in vitro* plants for phytoremediation studies can help to explore the enzymatic status and the products of metabolism of the dye, thus providing a new dimension to phytoremediation studies. The use of transgenic plants with microbial genes can combine the advantages of both plant and microbial systems for enhanced dye degradation. Biotechnological approaches involving the development of hairy roots and suspension cultures may find good utility in phytoremediation studies. The ultimate aim of phytoremediation involves applying these well studied plant systems at the contaminated sites which may constitute the development of constructed wetlands for on-site treatment of industrial effluents.

Chapter 1

INTRODUCTION

Since pre historic times innumerable uses of plants as sources of food, shelter, fuel etc., have been known to mankind. But, the newer approach of phytoremediation which involves the use of plant systems and/or rhizospheric organisms to remove content, inactivate or degrade harmful environmental contaminants and to revitalize contaminated sites (Vangronsveld et al, 2009), is an upcoming research area in the field of environmental biotechnology. Conventional techniques for bioremediation that involve the digging up of contaminated soils and disposal of the wastes to a landfill, lead to contamination elsewhere and can create significant risks in excavation, handling and transport of hazardous materials (Vidali, 2001). The chemical treatment methods used have multiple disadvantages such as their high cost, coupled with the formation of a large amount of sludge and the emission of toxic substances (Senan and Abraham, 2004), because of which bioremediation methodologies can be used as alternative technologies for the removal of industrial wastes. Microbial bioremediation processes for the removal of hazardous compounds, have received quite a lot of focus from researchers all over the world because of the high potentiality of prokaryotic systems to perform a variety of functions. But, the use of phytoremediation processes for the removal toxicants (especially textile dyes) is comparatively an unexplored methodology since the fact that plants also possess some inherent metabolic pathways that can breakdown a wide range of toxicants (Chaudhry et al, 2005) was much less realized. Since researchers have now begun to realize the potential of plant systems as effective remediating agents, this new area of phytoremediation has started gaining importance from academicia and industry (Cluis, 2004). Since plants are autotrophic systems of large biomass and require little nutrient input, phytoremediation technologies are easier to manage than microbial bioremediation systems and offer cost effective and aesthetically appealing options for environmental clean up (Cluis, 2004). Afforestation is one of the prescribed ways for minimizing the green house gases in the environment and reducing the effects of global warming since, plants have been known for their consumption of CO_2 and more recently of other gaseous industrial by products. Therefore, the value of plants to counterbalance the hazards of industrialization processes is being appreciated (Cumnningham and Ow, 1996). Phytoremediation can thus serve dual purposes.

The release of large amount of toxic wastes into water bodies is one of the consequences of increasing urbanization and industrialization in the modern world. A variety of organic (pesticides, explosives such as TNT, petrochemicals, chlorinated solvents, etc.) and inorganic (radionuclides, heavy metals such as mercury, lead, etc.) wastes which have toxic effects on the ecosystem have been contaminating our natural resources (Cluis, 2004). Out of the different types of pollutants released, dyes which are released by textile, dyestuff and dyeing industries constitute one recalcitrant group and are known to have carcinogenic and mutagenic effects with a potential toxicity to all life forms (Bafana et al, 2009). Most of the research involving phytoremediation technologies has been focused on the removal of heavy metals and a few organic compounds such as pesticides, polycyclic aromatic hydrocarbons etc. from the environment. The removal of textile dyes mediated by plant systems is still a much unexplored area of phytoremediation research. Hence, the article aims at reviewing the basic research and mechanisms involving the removal of dyes by plants and the application of these technologies at the dye contaminated sites with an insight into the future perspectives of research in this area.

Chapter 2

DYES-TOXICITY AND NEED FOR PHYTOREMEDIATION

Dyes are known to have complex structures that are difficult to degrade (Nilratnisakorn et al, 2007). With the advancement of technologies, enhancement has been made in dye properties so that they provide resistance to fading, provide improved delivery to fabrics and have increased variety of shades. These additional properties make them highly resistant to environmental degradation, thus increasing pollution (Togo et al, 2008). Sulfonated anthraquinones are generally the parent compounds for a vast array of dyes and thus the waste waters of textile industries are likely to contain these compounds which are recalcitrant and toxic (Page and Schwitzguébel, 2009). The difference in the chromophoric groups of dyes facilitates their classification into different types such as azo, triphenylmethane, anthraquinone, indophenol, diazonium, quinone dyes etc. Moreover, the nature of substituents attached to the basic aromatic ring structure also differs because of which they are not uniformly susceptible to bioremediation (Aubert and Schwitzguébel, 2004). In case of sulfonated dyes, the organosulfonate group plays an important role in altering the solubility and dispersion properties of the xenobiotic molecule and increases its recalcitrance to environmental breakdown, because of the thermodynamically stable carbon-sulfur bond (Duc et al, 1999). The mechanisms for the carcinogenicity of azo dyes that have been identified include metabolic activation to reactive electrophilic intermediates that covalently bind to DNA. Triphenylmethane dyes are known to cause reproductive abnormalities in rabbits and fish (Chen et al, 2010). Sometimes the products formed after the processing of these dyes themselves are toxic. In the environment, azo bonds of these dyes are reduced to liberate benzidine and other aromatic amines, which may cause adverse systemic health effects or cancer. Urinary bladder cancer is the most common form of cancer caused by exposure to benzidine. Stomach, kidneys, brain, mouth, esophagus, liver, and gall bladder might also be targets (Bafana et al, 2009). Toxicity of reactive dyes has been reported at concentrations as low as 5.2 mg/l (Nilratnisakorn et al, 2007). Hence, it is extremely important to implement technologies that completely remove such recalcitrant compounds from the environment or biotransform them into products that have reduced toxicity. Hence, for the targeted removal of such hazardous wastes from industrial effluents, plants can be used as efficient systems.

Chapter 3

Selection of Plant Systems for Remediation of Textile Dyes

For the removal of textile dyes from the environment, the selection of an appropriate plant with certain desirable characteristics is one of the most important preliminary steps in phytoremediation research. Though several plants have shown the ability to remediate contaminated soils; non edible plants are generally selected to be applied onto dye contaminated sites. Most of the studies on phytoremediation of textile dyes demonstrate their removal through either degradation of the dye or the adsorption and/or accumulation of the dye. Accumulation of organic compounds such as sulfonoaromatics has been shown in Rhubarb species (Duc et al, 1999). Compounds accumulated in the plant roots could be further translocated to shoots and leaves. To prevent the accumulated dye compounds or their metabolites from entering the food chain, the use of non edible plants is always preferred. Different types of grasses, ferns, weeds or agricultural wastes have been suggested and tested for the removal of dyes. *Phragmites* species have shown immense potential to remediate textile waste waters. *Phragmites australis,* a reed which is a component of the wetland community has been extensively studied for remediation of textile effluents and mainly with respect to the removal of the dye, Acid Orange 7 (Carias et al, 2007). Many native populations of *Phragmites autralis* are benign in that they pose little or no threat to other species. Among 10 different species of macrophytes screened, *Phragmites karka* was found to have broad amplitude of pH tolerance and was found to be growing well in alkaline, neutral and acidic textile wastewaters resulting in considerable shoot density and biomass to achieve maximum translocation of water and assimilation of nutrients. This makes the plants highly suitable for the treatment of textile waste waters that may be contaminated with different types of acidic as well as basic dyes. The good growth of underground organs in these species thereby provides maximum surface area to assimilate pollutants (Sharma et al, 2005). In addition, the plant should be fast growing and should have a deep rooted system that enables it to reach the pollutants easily. Larger biomass and surface area of the plant system can facilitate more efficient removal of the dye. Kagalkar et al have demonstrated that increase in plant biomass in terms of increasing number of plants used, gave higher % decolorization values. Reports have shown the efficient degradation of the dye Direct Red 5B (DR5B) with *Blumea malcolmii,* a deep rooted and fast growing plant system that forms sufficiently large biomass and can grow in soils with little nutrient availabilities (Kagalkar et al, 2009). Thus, a plant that has the ability to remove dye molecules from the environment and also have a good biomass can prove to be potent for phytoremediation. The plant *Typhonium flagelliforme* which has recently been reported for degradation of the dye Brilliant Blue R, exhibits dye degradation capacity when used even in distilled water, devoid of any other nutrients. The use of such plants can help to reduce the overall cost of the experiments (Kagalkar et al, 2010). Not all plants will be able to demonstrate similar responses or have similar removal rate for all dyes. In addition, for the phytotreatment of textile dyes, additional perspectives that should be kept in mind while selecting the plant species include high uptake rate of the pollutant, high translocation factor (TF) in case of phytoextraction, presenting the ability to translocate contaminants to the shoot and high tolerance level towards the dye (Zabłudowska et al, 2009). Hence, extensive screening of different plant species can help us to understand the selective abilities of a particular plant to remove a dye or a group of dyes. Moreover, textile effluents are generally mixtures of different dyes because of which plants that can be potent phytoremediators of textile industry wastes will be the ones that will be able to demonstrate the capacity to remove a large number and a diverse group of textile dyes eg., the species *B. malcolmii* showed the capacity to decolorize five different dyes, namely Direct Red 5B, Reactive Red 2, Methyl Orange, Malachite Green and Golden Yellow HER to varying extents (Kagalkar et al, 2009). *Typhonium* plantlets that have been studied for their phytoremediation potentialities, showed the additional advantage of remediating textile effluents and synthetic mixture of dyes along with 8 individual dyes, out of which maximum decolorization was obtained for the dye Brilliant Blue R (BBR) which was about 80%. To quantitate the % removal of color from dye effluents or synthetic dye mixtures, American Dye Manufacturers' Institute (ADMI 3WL) tristimulus filter method is used. The % removal of ADMI in case of mixture of dyes was 47% while in case of textile effluents was found to be 28% (Kagalkar et al, 2010). Many edible plants have been known to possess dye decolorizing abilities because of the rich enzymatic status of these plants. Though such plants are unsuitable for field applications, their enzymes can be extracted and used for degradation of various textile dyes. The species of *Sorghum vulgare*, *Phaseolous mungo* and *Brassica juncea* have shown the potential to decolorize the dye Reactive Red 2 and have also demonstrated to possess abilities to decolorize and detoxify textile effluents (Ghodake et al, 2009). Even though different plants are capable of degrading the same dye molecule, the products formed after degradation are likely to be different indicating that the pattern of transformation of the xenobiotic molecule is dependent upon the plant species (Page and Schwitzguébel, 2009). Plants that degrade the dye molecules into non toxic products are preferable for phytoremediation. Hence, the selection of the plant will depend upon the genetic make up of the plant that manifests in terms of varied enzyme activities in the plant and differential absorptive capacities resulting into variable patterns for the removal of dyes. Moreover, all the plants selected should be in the same stage of growth and should have almost equivalent dry weights and should have almost similar root and shoot lengths (Kagalkar et al, 2009) that can help to achieve reproducibility of results. Further, the plants selected should preferably be from the same area since factors such as age of the plant, soil conditions, nutrient status, light availability etc are factors that can affect the removal of the dye. In addition, the use of flowering plants for the removal of textile dyes would offer aesthetically appealing systems and will serve dual purposes of bioremediation and will also allow the flowers to be used for decorative purposes, thus serving economic benefits.

Chapter 4

PLANT MECHANISMS FOR THE REMOVAL OF DYES

A) MECHANISMS INVOLVING ADSORPTION AND/OR ACCUMULATION OF TEXTILE DYES

Plant mechanisms behind the removal of textile dyes, which are a group of organic pollutants, may be diverse. Though phytodegradation or phytotransformation are the most predominantly observed mechanisms adopted by plants for the degradation of organic compounds, the removal of textile dyes by plants also utilizes the mechanisms of adsorption and accumulation on plant surfaces. It has been established that the binding of xenobiotics to roots occurs by adsorption followed by its absorption into the plants (Davies at al, 2005). *Posidonia oceanica* leaf sheaths have shown effective adsorptive removal of the textile dye, Reactive Red 228. Moreover, the authors also demonstrated changes in the adsorptive capacities with changes in factors such as temperature and pH. Increase in temperature favored the better adsorptive removal of the dye which was probably because of the greater movement of the adsorbent material. Besides, the highest dye removal efficiency was found at pH 5 which might correspond to the rate of dissociation of the studied dye with maximum ionization of the molecule. The use of orange peels, banana peels, neem leafs, peanut hulls and agricultural wastes have also been suggested for the removal of various dyes (Ncibi et al, 2007). Untreated pulverized plant leaves of *Salsola vermiculata* revealed interesting adsorptive properties for Methylene Blue and iodine from aqueous solutions. Further, the activation of these plant leaves by treatment with zinc chloride gave better adsorptive properties. In this plant too, the effect of pH on the adsorption capacity of the chemically activated plant was investigated which showed a decrease in the adsorption capacity at lower pH values which may be because of the competition of protons with the dye molecules for available adsorption sites. The iodine number value which determines the capacity of the adsorbent to remove color from the solution, evaluated in terms of adsorption of iodine from the adsorbent pointed out that significant additional surface area can be achieved through zinc chloride activation and that microporosity contributes considerably to the total surface area of the prepared material making it a very good adsorbent for small compounds (Bestani et al, 2008). Thus, the inherent adsorptive capacity of the plant material can be enhanced through such activation techniques. The significant adsorption of Methylene Blue by this plant shows its potentiality in the remediation of textile dyes and effluents. Adsorptive removal of the dye Malachite Green by the roots of *Blumea malcolmii* was also reported where the % adsorption of the dye by the plant was approximately 45% (Kagalkar et al, 2009). The adsorption of a number of dyes was reported on the roots of *Typhonium flagelliforme* plantlets that were used for dye degradation experiments (Kagalkar et al, 2010). Phytoremediation processes not only involve the adsorption of the dye on the root or shoot systems of the plant but also involve the accumulation of dyes into plant tissues. Narrow-leaved Cattails (*Typha angustifolia* Linn.) that has the capacity to absorb a large amount of nutrients, has been demonstrated for the removal of the commercial diazo reactive dye, Reactive Red 141. The plant demonstrates the ability of 60% removal of the dye. After 28 days of exposure of the plant to synthetic reactive waste water containing the dye, the intercellular space of the plant showed the presence of the dye which was confirmed with transmission electron microscopy connected with electron dispersive X ray spectroscopy (TEM-EDX). Moreover, the plant also shows the precipitation of metal complexes with the dye which according to the authors are probably mechanisms to avoid damage to the plant (Nilratnisakorn et al, 2007). Rhubarb (*Rheum rabarbarum*) species have shown to accumulate synthetic anthraquinones which are starting materials for the production of a large number of synthetic dyes. The transpiration stream concentration factor (TSCF) was determined to detect the concentration of the accumulated compound in the xylem sap of these plants (Aubert and Schwitzguébel, 2004). When this value exceeds unity, the movement of the compound is faster than water. Among the different sulfonated anthraquinones the TSCF value obtained for anthraquinone-1-sulfonic acid was 2.5 which helped the authors to conclude that the movement of the pollutant was faster than water.

Anthraquinone-2-sulfonic acid and anthraquinone-2,6-disulfonic acid also showed values higher than unity. Plant screening showed that *Rheum rabarbarum* and *Rumex hydrolapatum* were most efficient for the accumulation of the five selected sulphonated anthraquinones (determined by the transpiration stream coefficient factor) which are precursors for many different dyes (Aubert and Schwitzguébel, 2004).

B) Plant Stress Response and Mechanisms for the Degradation of Dyes

Though adsorption and accumulation of dyes are important ways of phytoremediation, these processes lead to the mere concentration of pollutants from textile effluents onto and/or into plant surfaces and do not lead to the complete eradication of the pollutant. Thus, the phytotransforming abilities of a plant are of a greater significance since they can be employed to either completely degrade the dye or to transform it into products which are non toxic and can be safely released into the ecosystem. Because plants are static and live in a competitive and sometimes hostile environment, they have evolved mechanisms that protect them from environmental abiotic stress, including the detoxification of xenobiotic compounds (Page and Schwitzguébel, 2009). Plant mechanism is diverse and can be used to treat compounds not degradable by bacteria. Different aromatic compounds such as derivatives of sulfonated anthraquinones occur naturally in several plant genera and thus these plants are likely to possess enzymes that can accept these aromatic compounds as substrates and process them (Aubert and Schwitzguébel, 2004). An important step in the removal of sulfonated anthraquinones appears to be involving the action of dioxygenases adding oxygen across the double bond bearing the sulfonate group leading to its elimination (Schwitzguébel et al., 2002). Plant degradation of textile dyes may either be intracellular involving enzyme systems inside plant tissues or it may involve degradation with the help of extracellular enzymes secreted by the plant in rhizosphere regions. The hydrophobicity of a compound can affect its uptake or translocation. Moderately hydrophobic dyes can be most readily taken up by the plant or translocated within the plant. Hydrophobic compounds can also be bound to root surfaces or partition into roots but cannot be further translocated into the plant (EPA, 2000). Thus, phytodegradation of the dye outside the plant will not depend upon plant uptake.

Plant detoxification pathways comprise of three phases with specific enzymes. Phase I enzymes like cytochrome P450 and peroxidases transform xenobiotics (mainly by oxidation reactions) in order to allow the conjugation of the oxidized xenobiotic with glutathione catalyzed by glutathione S-transferase, in phase II. Phase III involves the translocation of these conjugates into vacuoles (Carias et al, 2008). The degradation process of dyes has shown to involve a significant role of peroxidases that are enzymes which are typically activated as an enzymatic stress response and are comparatively more extensively studied in plants than the other enzymes that can have a role in textile dye degradation. This strategy where peroxidases are activated as stress response appears to be very interesting as plants not only allow the pretreatment of specific recalcitrant compounds by changing their physicochemical properties and making them more amenable for treatment but also in their transformation into innocuous products (Carias, 2008). Lignin peroxidases (LiP) are the primary enzymes that are involved in the lignoloysis of wood and they oxidize lignin structures by one electron yielding cation radical intermediates that undergo spontaneous fission reactions. Studies with lignin model compounds have shown that LiP cleaves the predominant aryl glycerol β-aryl ether substructure of lignin which accounts for about half the total polymer, between Cα and Cβ of its propyl side chain (Sarkanen et al, 1991). They are known to act upon a variety of xenobiotic compounds and mediate the symmetric or asymmetric cleavage of many dyes at their C-C linkages. Thus, this enzyme is found to be one of the most predominant enzymes involved in the transformation and/or degradation of dyes. Peroxidases are heme containing enzymes able to oxidize a wide range of organic and inorganic compounds, using hydrogen peroxide as a co-substrate. They are non specific and can use a broad range of electron donor substrates. The first step of action of this enzyme leads to the cleavage of hydrogen peroxide molecule with the concomitant production of water and incorporation of one of the oxygen atoms of hydrogen peroxide into the initial compound. The next two steps involve the reduction of enzyme in order to regenerate it. Pollutants like azo dyes act as electron donor substrates (Davies et al, 2005). Many plant species have shown the presence of peroxidases in their tissues. The involvement of these enzymes was shown in curly dock plant that was used for the decolorization of the dye Remazol Brilliant Blue R where it was found that addition of hydrogen peroxide into the cultivation medium of the plant greatly stimulated the decolorization of this dye (Takashi et al, 2005). Similar observations were noted in case of degradation of the dye Acid Orange 7 by *Phragmites australis*. On the addition of H_2O_2 at 192 h of exposure of the plant to the dye, a significant reduction in the absorbance of the dye was found after 48 h and 120 h of H_2O_2 addition (Davies et al, 2005). As a response to the polydye R-478, the MPH-4 clonal lines of *Mentha pulegium* showed an increase in the guaicol peroxidase activity while decrease in the phenolic content which probably indicates that the phenolics had been used up for the lignin biosynthesis process (Strycharz and Shetty, 2002a), which could be stimulated because of the increased peroxidase activities. Similar results with phenolic content and guaicol peroxidase activity were demonstrated in case of oregano (*Origanum vulgare* L.) cell lines (Strycharz and Shetty, 2002b). In these studies, the authors speculate that the dye might be used as a substrate for peroxidase cross linking and might participate in lignification process though there is no evidence provided. An induction in intracellular peroxidase activities in the roots was also found during the degradation of the dye DR5B, mediated by *Blumea malcolmii* after 3 days of exposure to the dye when the plant showed 60% decolorization of the dye. The role of peroxidase in the degradation of DR5B, mediated by *Blumea,* can be predicted in catalyzing the asymmetric C-C cleavage of the intermediates formed during processing of the dye by the plant enzyme systems (Kagalkar et al, 2009). The root tissues of the plant *Typhonium flagelliforme* also showed induction of intracellular and extracellular peroxidase values upon exposure to the dye Brillaint Blue R (Kagalkar et al, 2010). The exposure of the dye Reactive Red 198, to *Tagetes patula* L. hairy roots also demonstrated a significant induction in peroxidase activities. Both *Typhonium* and marigold peroxidases can also be predicted to be having a similar role in the degradation of Brilliant Blue R and Reactive Red 198 respectively, where the enzyme probably catalyzes the asymmetric cleavage of the original dye molecule itself (Patil et al, 2009; Kagalkar et al, 2010). The presence of oxidoreductive enzymes was assessed in three types of plants, Alfalfa, Mustard and Cresswhich which showed that peroxidases were the most dominating enzyme species found in the root, shoot and exudates of these plants (Gramss and Rudeschko, 1998).

Cytochrome P450 monooxygenases represent a multigenic family of enzymes, involved in the detoxification process of many xenobiotic compounds. In addition to activating xenobiotics, cytochrome P450 plays an important role in the normal secondary metabolism of plants, which produce compounds involved in cell signaling and defense mechanisms. A significant activity of the enzymes was detected in the leaves of Rhubarb with different anthraquinones as substrates. Lower activities of these enzymes were also detected in roots and petioles. In contrast, when the authors used common sorrel no significant difference was found in the activities of the plants which were exposed and those which were unexposed to the anthraquinones. The higher activities of the enzyme in the leaf tissues of Rhubarb indicated that the sulfonated anthraquinones were taken up by the plant and were translocated in the leaf tissues (Page and Schwitzguébel, 2009). The ability of these plants to accumulate sulfonated anthraquinones has been confirmed by the studies performed by Aubert and Schwitzguébel (Aubert and Schwitzguébel, 2004). This assumption was supported by the data obtained with capillary electrophoresis which confirms the presence of anthraquinones in leaves of Rhubarb plants (Aubert and Schwitzguébel, 2002). Further, new metabolites were also found to be present in leaf tissues of plants exposed to the anthraquinones which indicate the biotransformation of these compounds, but the metabolites have not yet been identified (Aubert and Schwitzguébel, 2004). These results indicate that Rhubarb probably has the ability to not only accumulate sulfonated anthraquinones but also biotransform them.

Another detoxifying enzyme that has been studied in plants is glutathione S-transferase. The best known role of this enzyme is in the detoxification of endobiotic and nucleophilic xenobiotic compounds by covalently linking GSH to a broad variety of reactive electrophilic and hydrophobic substrates, which results in the formation of more polar and less reactive conjugates. An increase in the activities of this enzyme has been studied in *Phragmites australis* in response to Acid Orange 7 (Carias et al, 2008). The plant has shown the potential to degrade the dye with a removal efficiency of $68 \pm 8\%$ when used in constructed wetlands. The presence of a dye in the vicinity of a plant can offer stress conditions which the plant will try to overcome with inherent stress response mechanisms. Plants react to stress conditions by increasing the concentration of reactive oxygen species using enzymes like NADPH oxidase in order to signal plant defences. This in turn leads to the activation of antioxidant scavenging enzymes to remove the reactive oxygen species. Superoxide dismutase (SOD) converts the reactive oxygen species into hydrogen peroxide which will then be converted into water and oxygen, by the action of peroxidase, catalase and ascorbate peroxidase. An induction in the activities of these antioxidative and detoxifying enzymes was observed following exposure to the dye and these enzymes may play an active role in the degradation mechanism of this dye. The authors speculated that this induction could be attributed either to the activation of the antioxidant scavenging enzyme battery or could also be due to the *de novo* synthesis of enzymatic proteins to face the stress conditions. Molecular studies performed show that Acid Orange 7 acts as a chemical stressor agent for *Phragmites australis,* activating the gene expression for

Cu/Zn SOD, Mn SOD, glutathione peroxidase and catalase isoforms, which leads to the conclusion that this gene over expression is related with the sudden production of reactive oxygen species after exposure to high concentrations of the dye Acid Orange 7 (Davies et al, 2009). Increase in the content of dehydroascorbate reductase which is involved in the regulation of ascorbate glutathione pathway was also observed in *Phragmites* plants exposed to the dye. The dye seems to be incorporated in the cytosol of the plants, where via the detoxification pathway; it is modified (Phase I), conjugated (Phase II) and translocated (Phase III) into the vacuoles. Increase in the activity of glutathione S-transferase indicates that the dye is being compartmentalized in the cells (Carias et al, 2008).

Laccases constitute another major class of degradative enzymes that have been studied. They constitute one class of polyphenol oxidases that catalyze the oxidation of various substituted phenolic compounds by using molecular oxygen as an electron acceptor. They catalyze the removal of hydrogen atom from the hydroxyl group of *ortho-* and *para*-substituted mono and polyphenolic substrates and from aromatic amines by one-electron abstraction to form free radicals capable of undergoing further depolymerization, repolymerization, demethylation or quinone formation (Abadulla et al., 2000). The inherent ability of plant systems to produce these enzymes probably stems from the role of these enzymes in plant development and lignification. Though laccases have been found in various plants such as peach, sycamore, tobacco and poplar, they have not been characterized or used extensively because their detection and purification is often difficult as crude plant extracts often contain a large number of oxidative enzymes with broad substrate specificities (Sharma et al, 2007). Laccases can have immense potential in the detoxification and degradation of wastes form textile industries. Fungal and bacterial laccases have been well characterized and their role in bacterial and fungal dye decolorization has been well illustrated. Though studies on the involvement of plant laccases in phytoremediation of dyes are few, recently Ghodake et al reported the significant induction of intracellular laccase in roots of *Brassica juncea* upon exposure to the dye Reactive Red 2 and textile effluent (Ghodake et al, 2009). Laccase was also found to be induced in marigold hairy roots during decolorization of the dye Reactive Red 198 (Patil et al, 2009). An interesting feature of these studies indicated the absence of this enzyme in roots unexposed to the dye.

Tyrosinases are also a group of copper containing polyphenol oxidases that catalyze two type of reactions, the *o*-hydroxylation of some monophenols (monophenolase, cresolase) and the oxidation of *o*-diphenols to *o*-quinones (diphenolase, catecholase) using molecular oxygen (Chen and Flurkey, 2002). These enzymes have been studied for their role in microbial dye degradation but dye degradation mechanisms associated with plants have been poorly studied in context with these enzymes. The presence of the dye Direct Red 5B, in the medium has shown an induction in intracellular tyrosinase activities while the presence of Reactive Red 198 in the medium has resulted into an induction in both the intracellular and extracellular activities of the enzyme which indicates the vital role of these enzymes in dye degradation processes (Kagalkar et al, 2009; Patil et al, 2009).

The other enzymes that were found to be induced in *Blumea malcolmii* during the degradation of Direct Red 5B include riboflavin reductase, azoreductase and NADH-DCIP reductase which indicates their involvement in dye degradation processes. Azoreductase is known to catalyze the break down of azo linkages in azo dye structures. Direct Red 5B being an azo dye can be predicted to be acted upon by azoreductase from *Blumea* species leading to the degradation of the dye into simpler molecules (Kagalkar et al, 2009). Enzymes such as laccase, catechol 2,3-dioxygenase, ascorbate oxidase have also been detected in many other plants such as Alfalfa, Mustard and Cresswhich but, their application for dye removal has not been studied in these plants (Gramss and Rudeschko, 1998). Thus, the enzymatic status and the mechanisms of dye removal are found to be highly variable with different plant species. An evidence for this is given by the studies on two different plant systems, *Blumea malcolmii* and *Typhonium flagelliforme* which even upon exposure of the same dye molecule exhibited different enzyme activity patterns. Azoreductase activity was found to be absent in *Typhonium* species whereas the presence and induction in its activity was reported in *Blumea*, when both the plants were exposed to the same dye, DR5B. Similarly intracellular laccase was found to be present in *Typhonium* root tissues but absent in *Blumea* roots. The activities of riboflavin reductase which were found in *Blumea* species after exposure to Direct Red 5B was not detected in *Typhonium* species exposed to the same dye molecule. Moreover, the enzyme activities in the same species can be found to vary upon exposure to two different dyes. Enzyme activity in *Typhonium flagelliforme* plantlets exposed to two different dyes, DR5B and BBR shows higher induction in the values of peroxidase and laccase in case of plantlets exposed to BBR than DR5B, which according to the authors could be the reason behind the better removal of BBR by *Typhonium* than DR5B. *Blumea* plantlets exhibited induction in azoreductase activities on exposure to DR5B, while no activity of the enzyme was detected for BBR (Kagalkar et al, 2009; Kagalkar et al, 2010).

Since reports suggest that plant species are capable of adsorbing as well as accumulating dyes, it is of great importance to confirm that the decolorization of the dye is not because of the adsorption or accumulation of the dye onto and/or into plant species but it is because of the transformation or degradation of the dye into different products. HPLC analyses showing the different retention times observed for peaks of the individual dye and the extracted products formed after the treatment of the dye, with the selected plant species, help to confirm the degradation of the dye. Simpler techniques such as thin layer chromatography can also be used for such analyses. The studies carried out on the plants *B. juncea*, *B. malcolmii* (Figure 3) and marigold hairy roots have confirmed the degradation of the dye, detected through techniques such as high performance liquid chromatography (Ghodake et al, 2009; Kagalkar et al, 2009; Patil et al, 2009).

Chapter 5

ANALYSIS OF PRODUCTS FORMED AFTER THE PHYTODEGRADATION OF DYES

A) ANALYSIS OF PRODUCTS WITH RESPECT TO THEIR CHEMICAL STRUCTURES

It is indeed very important to predict the chemical nature of the degradation products which can be done by correlating the analysis of gas chromatography mass spectroscopy (GCMS) techniques with fourrier transfrom infrared spectroscopy (FTIR). The FTIR spectra obtained helps us to predict the changes occurring in the functional groups of the original dye molecules, after degradation by the plant system. FTIR spectral data can further help to give confirmatory evidences in favor of the formation of products that have been predicted as a result of the degradation of the dye, by GCMS techniques. The probable products formed after the degradation of Direct Red 5B by *Blumea* species have been predicted to be 4-(4-amino-phenylazo)-benzene sulfonic acid, 3-amino-7-carboxyamino-4-hydroxy-naphthalene-2-sulfonic acid and 7-carboxyamino-naphthalene-2-sulfonic acid (Figure 1) (Kagalkar et al, 2009). Sometimes, the exudates secreted by the plant can be misinterpreted as metabolites formed after the degradation of the dye, thus giving false positive results with high performance liquid chromatography (HPLC), FTIR and GCMS techniques. Hence it is indeed very important to run controls that will help us to clearly differentiate between the products formed after the degradation of the dye molecule and those compounds which are normally secreted as plant exudates.

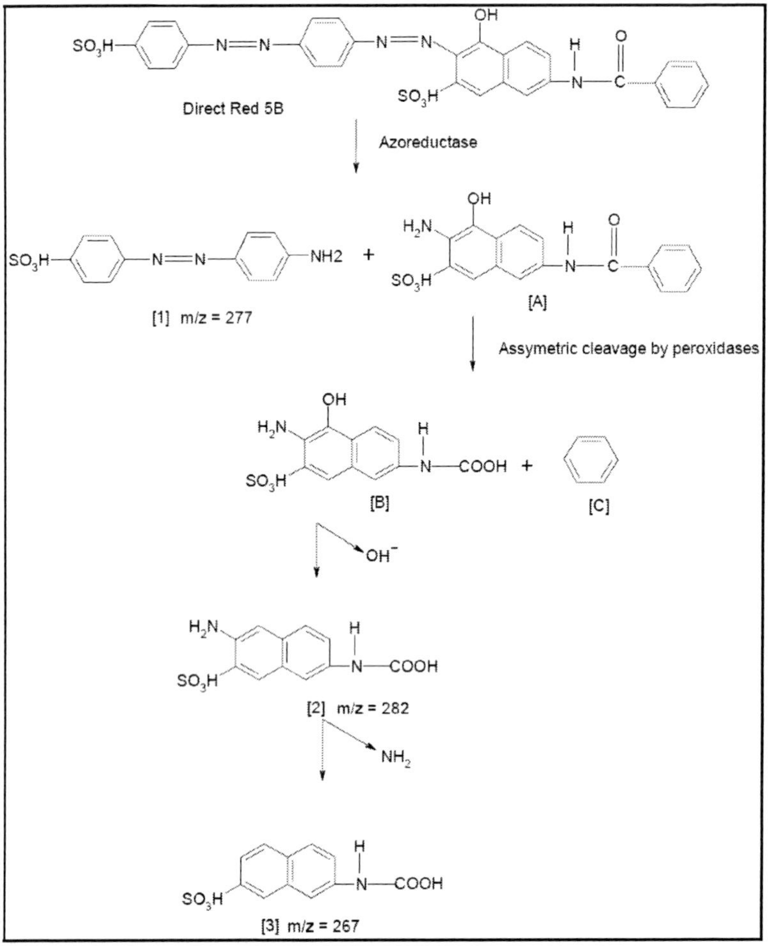

Figure 1. Proposed pathway for the phytotransformation of the dye Direct Red 5B by *B. malcolmii.* The compounds represented by alphabets have not been found, but their existence is rationalized as necessary intermediates for the final products found. The compounds in Arabic numbers have been found in reaction mixture (Kagalkar et al, 2009).

The FTIR and HPLC analyses of three different samples including the dye Direct Red 5B, the exudates of the plants along with the dye and the degraded sample clearly helped to select peaks that were neither of the plant exudates nor of the original dye molecules and thus could be easily concluded to be the peaks representing the formation of new products owing to the metabolism of the dye (Figure 2 and 3) (Kagalkar et al, 2009).

Similarly, the products formed after the degradation of the dye Reactive Red 2 by *B. juncea* species were found to be naphthalene sufamide and 2-amino-4, 6-dichlorotriazine while the products formed after the degradation of Reactive Red 198 by marigold hairy roots were predicted to be 2-aminonaphthol, *p*-aminovinylsulfone ethyl disulfate and 1-aminotriazine, 3-pyridine sulfonic acid (Ghodake et al, 2009; Patil et al, 2009). The identification of these chemical structures also helps to portray the role played by different degradative enzymes in the sequential metabolism of dye structures that can help us to design the probable pathway of metabolism of the dye. Shaffiq et al also analyzed the products formed after the degradation of various dyes by *Ipomoea* and *Saccharum* peroxidases by HPLC techniques and concluded that the products were not aromatic amines which are known to be toxic (Shaffiq et al, 2002).

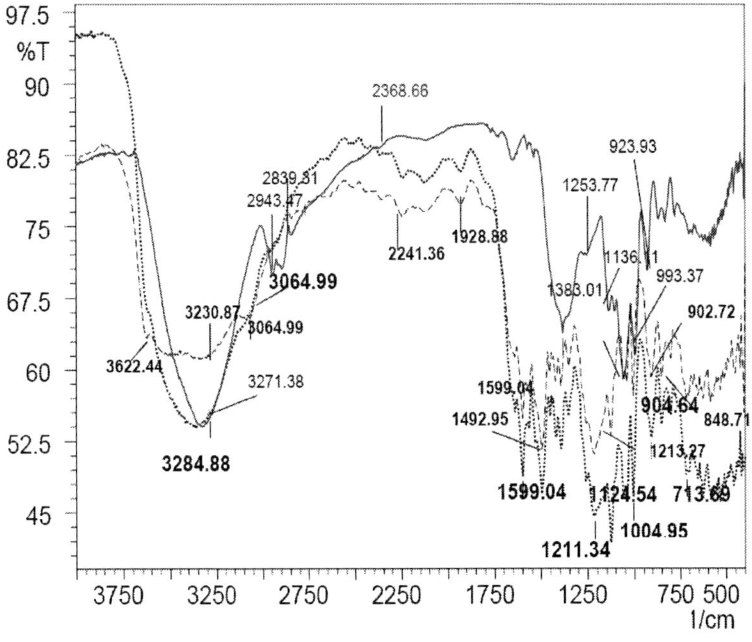

Figire. 2. FTIR spectral analysis of Direct Red 5B (---), extracts of *Blumea malcolmii* exudates along with the dye (···) and products formed after the degradation of the dye by *Blumea malcolmii* (–) (Kagalkar et al, 2009).

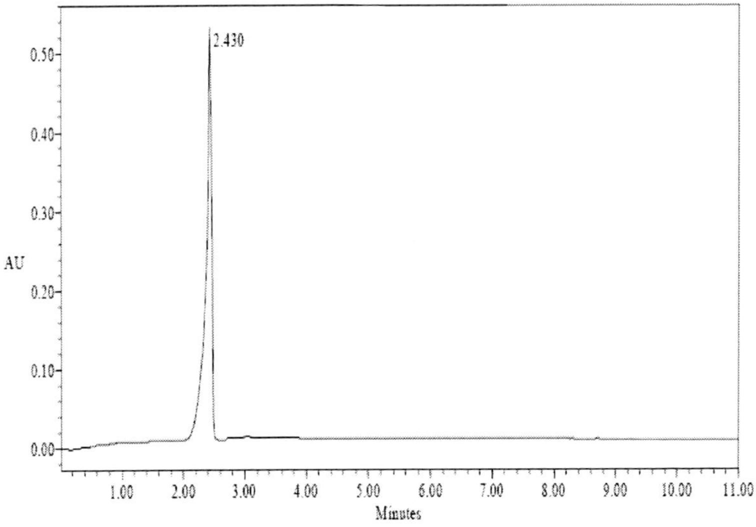
2.430
0.50
0.40
0.30
AU
0.20
0.10
0.00
1.00
2.00
3.00
4.00
5.00
6.00
7.00
8.00
9.00
10.00
11.00
Minutes

a

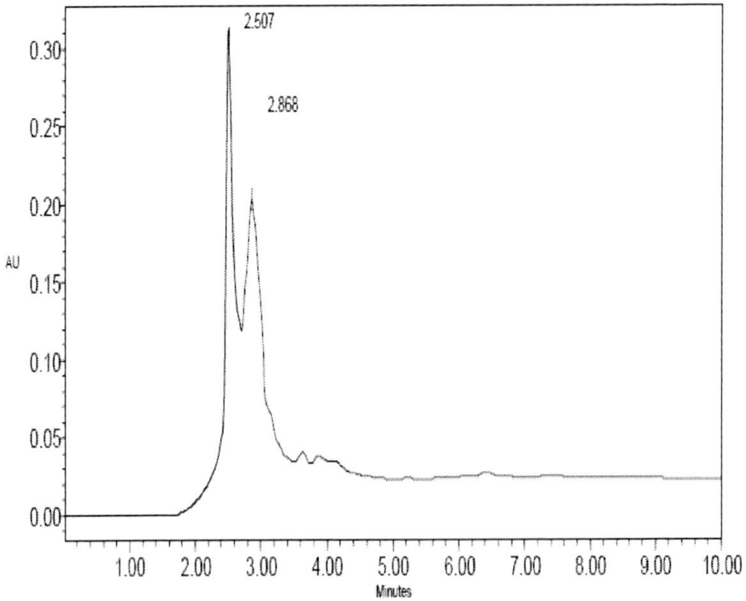
2.507
2.868
0.30
0.25
0.20
AU
0.15
0.10
0.05
0.00
1.00
2.00
3.00
4.00
5.00
6.00
7.00
8.00
9.00
10.00
Minutes

b

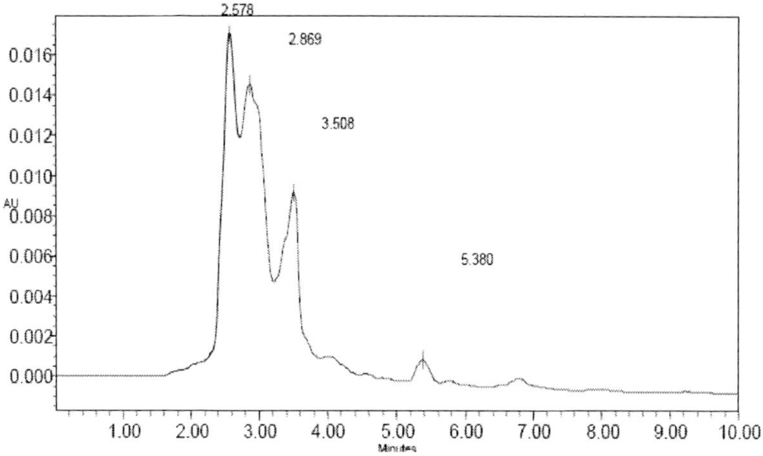

c

Figure 3 (a). HPLC analysis of the individual dye, Direct Red 5B. (b) HPLC analysis of the control sample containing extracts of exudates of the plant, *Blumea malcolmii* and Direct Red 5B. (c) HPLC analysis of the test sample containing degraded products of Direct Red 5B using *Blumea malcolmii* (Kagalkar et al, 2009).

Chapter 6

ANALYSIS OF PRODUCTS FORMED WITH RESPECT TO THEIR TOXICITY

Before applying any plant system at the contamination site, it is very important to have a detailed knowledge about the basic mechanisms underlying the removal of the dye by the plant species. Moreover, it is also very essential to determine the nature of the products formed with the degradation of the dye in terms of the chemical nature of the product and also its toxicity to different life forms including microorganisms, aquatic animals like fishes which are very frequently exposed to contaminated water supplies, plant systems etc. The phytotoxicity studies of the products formed after the degradation of Reactive Red 198 and BBR, after treatment with marigold hairy roots and *Typhonium* plantlets, towards the plants *Phaseolus mungo* L. and *Triticum aestivum* L. revealed the non toxic nature of the metabolites formed (Patil et al, 2009; Kagalkar et al, 2010). Such studies then indicate that the plant systems can be well applied in soil or water ecosystems and their application will only be beneficial to the ecosystem and will not lead to the formation of products that will be toxic for any life form. In addition, it is also important to analyze the toxicity of the dyes to be used towards the plants that have been selected for their application in dye removal. When narrow leaved cattails was subjected to high concentration of dyes from 100 to 300 mg/1, the plants responded by showing symptoms such as green wilting, then a yellow spot was observed which is a symptom of necrosis after 48 h of exposure to the dye. Thus, higher dye concentration was found to be toxic to the plant and at concentrations of 300 mg/1; there was no survival of the plant. The toxicity of the dye to the plant was found to be at and above the concentration of 25.33 mg/l. Such type of studies helped the authors to ascertain a dye concentration of 20 mg/l for further experiments (Nilratnisakorn et al, 2007). Kagalkar et al also reported the effect of increasing concentration of the dye Direct Red 5B on the percentage decolorization values. Increasing concentrations of the dye led to a decrease in the percentage decolorization values, indicating that higher dye concentrations could be toxic to the plant (Kagalkar et al, 2009). When the degradation of textile effluents or synthetic dye mixtures is studied the reduction BOD (Biological Oxygen Demand), COD (Chemical Oxygen Demand), TS (Total Solids) and TDS (Total Dissolved Solids) etc. also constitute important parameters in the assessment of toxicity of the waste waters. *Typhonium flagelliforme* plantlets reduced the BOD of the textile effluent as well as of the synthetic mixture of dyes. Similarly, the COD value of the industrial effluent was reduced (Kagalkar et al, 2010). Reduction in these values indicates the reduced toxicities of these effluents.

Chapter 7

THE USE OF HYDROPONICS AND PLANT TISSUE CULTURE TECHNOLOGIES FOR DYE DEGRADATION

The cultivation of plants and their further experiments with dye degradation can be carried out using hydroponic solutions. These solutions provide a nutrient status which is close to that of the soil in which the plant usually grows. Thus, such solutions are enriched with various macro and micro nutrients and can be used for the cultivation and/or maintenance of plants for phytoremediation. The use of hydroponics provides a cost effective method for phytoremediation of dyes. Aubert and Schwitzguébel carried out the screening of plant species (*Rheum rabarbarum, Rumex acetosa, Rumex hydrolapatum and Apium graveolens*), in hydroponic solutions for the removal of sulfonated anthraquinones. Many plant species have the capacity to absorb large quantities of water form hydroponic solutions. The water absorption capacity of a plant is a factor that should be taken into consideration while performing studies in hydroponic solutions because it reflects the overall health of a plant. Lower water absorption capacity for the plant *Rumex acetosa* in hydroponic solution indicated that the plant was not in optimum health under hydroponic conditions and thus the metabolism and transpiration was probably reduced as compared to soil grown plants. Though difficult, it is quite possible to grow adult terrestrial plants such as Rhubarb and common sorrel under hydroponic conditions (Aubert and Schwitzguébel, 2004). But, research that has been involving the cultivation and experimentation with plants in such systems also portrays some major disadvantages of these systems. Pege and Schwitzguébel found it impossible to collect leaves of the same age and same stage of growth and development in case of plants grown in hydroponics. It has been found that the level of enzymes like cytochrome P450 changes with the growth of the plants since they play a role in several physiological functions of the plant. The same is true with peroxidases. This methodology of work makes it difficult to exactly confirm the role of these enzymes in the detoxification of dyes (Pege and Schwitzguébel, 2009). Moreover, the enzyme activities of a plant may also be affected by conditions such as nutritional status, dark and shade requirements, effect of microbial contaminants etc. These factors make it very difficult to get reproducible results. To overcome these problems with wild plants grown in hydroponics or in wetlands, the importance of tissue culture based technologies has been stressed by a few researchers. Though tissue culture involves processes that require a high cost, the use of these technologies has been suggested for basic research which lays the foundation for the application of plant systems in wetland conditions. Our current ability to exploit phytoremediation technologies for the treatment of dyes is restricted by the fact that the knowledge regarding the basic mechanisms and pathways involving the decolorization of dyes is limited. The advantage of using tissue culture based technologies is that the plants can be grown in controlled conditions and once established, they can be propagated indefinitely and are available on demand as contrast to whole plants that have a limited life span. Moreover, *in vitro* culture techniques offer an environment that is totally free of microbial contamination and can be used to distinguish the responses and capabilities of plant cells from microbes present in the rhizhospheric regions in dye contaminated sites. They also offer conditions that are controlled in terms of nutrient levels, phytohormone level, light requirements etc. (Doran, 2009). Thus, Kagalkar et al have reported studies on the decolorization of various dyes using *in vitro* plants. These techniques have given reproducible results and have led the authors to analyze the role of different enzymes involved in the degradation of the dye DR5B and BBR (using the plants, *Blumea malcolmii* and *Typhonium flagelliforme* respectively), also predict the probable pathway behind the metabolism of these dyes (Kagalkar et al, 2009; Kagalkar et al, 2010). Tissue culture technologies also help to manipulate the plant in order to obtain callus cultures, suspension cultures and hairy roots. Hairy roots grow relatively quickly and do not require exogenous hormones in the medium (Doran, 2009). These advantages have led to the use of marigold hairy roots in the degradative analysis of the dye Reactive Red 198 (Patil et al, 2009). Experiments with separately cultured organs of a plant can evaluate the accumulation and/or biotransformation abilities thus minimizing the interference of translocation effect of dyes (Doran, 2009). Though plant tissue culture technologies offer multiple advantages, their use is feasible only for studying the basic processes and not for the actual applications of dye degradation in the field. Moreover, the characteristic features of the plants that are demonstrated in hydroponics or *in vitro* culture conditions could be different than those observed in the soil because of the complexity of the soil environment which exposes the plant to different biotic (microorganisms) and abiotic (other contaminants) soil elements (Zabłudowska et al, 2009). Hence such experiments should always be accompanied with field trials of the plant.

Chapter 8

Synergistic Approaches for Dye Degradation

Rhizhosphere remediation constitutes an interesting branch of phytoremediation technologies involving the use of plants along with rhizospheric microorganisms to remediate contaminated soils. In addition to the root zone (rhizosphere), where the microbial biomass can be one order of magnitude or more higher than that in bulk soil, bacteria can colonize the interior of their host plant without causing symptoms of disease (Weyens et al, 2009a). Researchers have shown that plants can be able to degrade a wide variety of organic pollutants in association with microbes (Peng et al, 2009). The limitations to remediation processes can be overcome by utilizing the dynamic synergy between plants and rhizhospheric organisms. Soil microorganisms are also known to produce certain biosurfactant compounds that may further facilitate the removal/degradation of organic pollutants by increasing their availability to plants. The genera commonly found in rhizospheres include *Rhizobium*, *Azotobacter* and *Pseudomonas* as well as a number of root-associated fungi, including those involved in the formation of symbiotic mycorrhizas (Chaudhry et al, 2005). Root exudates are compounds produced by the plant and released by the plant roots. These exudates contain water soluble, insoluble, and volatile compounds including sugars, amino acids, organic acids, nucleotides, flavonones, phenolic compounds and certain enzymes. The microbial populations and activity in the rhizosphere can be increased due to the presence of these exudates, and can result in increased organic contaminant biodegradation in the soil (EPA, 2000). In return many microbes have shown the potential to boost plant growth directly by secretion of different phytohormones or by fixing and solubilizing nutrients that are unavailable to plants and indirectly by competing with plant pathogens for availabilities of nutrients and space (Weyens et al, 2009b). Microbes have been known for their potential to degrade a variety of dyes with diverse structures and properties. There have been reports of the presence of different enzymes such as lignin peroxidase, laccase, tyrosinase, DCIP reductase, azoreductase etc in microbial species (Kalme et al, 2006). It is quite likely that the degradation of a dye mediated by individual microbial or plant systems could lead to partial degradation. Use of microbial and plant systems together with their diverse inherent enzyme producing abilities can thus be used together in an attempt to obtain the mineralization of the dye molecule and return to the nature its elements in their native forms. Strycharz et al showed that when oregano cell lines were inoculated with *Pseudomonas* Z strain, hyperhydricity was prevented, thus improving the quality of the plant. An increase in the activity of peroxidase was found when *Pseudomonas* was used in combination with the dye without a substantial decrease in the phenolics. This suggests that this *Pseudomonas* species may have the potential to degrade the dye. Moreover, the authors speculate that as many *Pseudomonas* species have azoreductase activities, they could break the azo linkages in the dye and convert it into a more biologically available form that can be degraded by the plant. If plant-bacterial combinations are to be used to the fullest advantage of remediating the soil, it will be good to establish tolerant clonal lines before planting on contaminated soil (Strycharz, 2002b). Research involving synergistic approaches for phytoremediation is still at the very initial stage which is probably because of the lack of knowledge behind the basic mechanistic processes of plant dye degradation.

Chapter 9

Applications of Phytoremediation Technologies for Dye Degradation

A) Constructed Wetlands

Most of the research involving the degradation of different pollutants is limited to laboratory conditions and very few of it is actually applied in the field. The use of constructed wetlands can take us a step closer to the application of potent plant species on the actual sites of contamination. Experiments performed in the laboratory are performed under controlled conditions and the behavior and efficiency of the system when applied at the actual site of contamination remains a question. Constructed wetlands are engineered systems to treat waste water (Barbera et al, 2009) and can be designed to mimic conditions close to those prevalent at the dye contaminated sites. Moreover, unlike natural wetlands, depuration processes in constructed wetlands are performed under environments which are more controlled and thus can assure greater efficiency and regular depuration activity across the entire bed (Barbera et al, 2009). Biodegradation of less-degradable pollutants generally requires combination of anaerobic and aerobic processes. For example, azo dyes, sixty to seventy percent of dyes used in the textile industry, are mineralized aerobically only after the azo-linkage is broken anaerobically (Ong et al., 2005). To treat such pollutants with constructed wetlands, therefore, anaerobic processes should properly incorporated to wetland systems. A vertical flow constructed wetland was designed so as to work in intermittent feeding mode (8 feeding cycles per day) which enhanced the characteristics like constant hydraulic permeability and maximized the oxygen transfer rate and was tested for the removal efficiency of the dye Acid Orange 7. The flow rate of the effluent must be critically maintained in such experiments. The use of sprinklers for feeding the dye solutions on the beds allows good distribution of the dye over the entire surface area (Davies et al, 2005). The main role of wetland vegetation is attributed to the modification of soil texture, hydraulic conductivity and soil chemistry. For phytoremediation technologies to be applied in constructed wetlands, parameters such as BOD, COD, TOC content, hardness, alkalinity etc of the effluent should be evaluated before and after treatment of the industrial effluent. The nature of the products formed should also be determined. The effluent treated by *Phragmites australis* plants in constructed wetlands showed an efficient reduction in the COD and TOC levels (Davies et al, 2005). Thus, constructed wetlands can be used as economic technologies that can treat enormous amounts of wastes through batch or continuous processes. The efficiency of a wetland system depends largely on the basic biological, physicochemical processes induced by the interaction of plants, microorganisms, substrates and pollutants (Barbera et al, 2009).

B) Purified Enzymes for Phytoremediation

Plants remain indispensable sources for the production of a number of biologically active chemical substances that are difficult to synthesize because of their complicated structures (Rudrappa et al, 2005). The uses of degradative enzymes that have been studied for dye colorization are seen as attractive options in the development of effective strategies for the biological treatment of waste water. The catalytic action of enzymes is highly specific and efficient as compared to chemical catalysis because of the higher reaction rates, milder reaction conditions and greater stereospecificity (Maddhinni et al, 2006). One of the major drawbacks of using these enzymes for remediation purposes is the low yield and high cost of production as compared to bacterial and fungal enzymes. These processes can be made economic either by reducing the production cost or by extracting these enzymes form cheaply available plant sources and increasing the purification fold and percentage of recovery after purification (Shaffiq et al, 2002). Enzymes can be immobilized on suitable carriers after purification and can be used for remediating dye effluents. Immobilization techniques offer several advantages such as easy separation from the soluble reaction products and the untreated substrate. Moreover, immobilization techniques allow repeated usage and can help to reduce the overall cost of the process.

They allow the continuous removal of toxic metabolites thus simplifying the work (Matto et al, 2009). The stability of carrier enzyme binding is an important factor in the application of immobilized enzymes. Moreover, matrices such as silica, polyvinyl alcohol, polyacrylamides etc, allow the adsorption of dye molecules on the matrix which may lead to the inactivation of the enzyme (Shaffiq et al, 2002). Further, it is very important that the enzyme retains its activity after immobilization. Once the enzyme is successfully immobilized it can be used in bioreactors. The plant systems, *Ipomoea palmata* and *Saccharum spontaneum* served as good sources of peroxidase. The enzyme purified from these sources was found to decolorize various dyes. But, the pH optima for degradation of acidic dyes by *Ipomoea* peroxidase was found to be between 4.0 and 6.0 while the pH optima for the degradation of basic dyes was found to be between 6.0 and 8.0. The dye Supranol Green (25 mg/l) was most efficiently degraded to about 84% within 4 h of treatment with the purified enzyme followed by Brilliant Green (25 mg/l) which showed 54% degradation for the same dye concentration. Similarly, the optimum pH for the decolorization of various dyes by *Saccharum* peroxidase was different. Supranol Green (25 mg/l) was again the most efficiently removed to get 99% degradation within just 20 min of reaction time with the enzyme. *Saccharum* peroxidase being more efficient in the degradation of dyes was immobilized using modified polyethylene which is a hydrophobic matrix and thus prevents the adsorption of dyes onto the matrix. The immobilized enzyme retained 20% of its activity and could efficiently decolorize Procion Green HE-4BD, Supranol Green, Procion Brilliant Blue H-7G and Procion Navy Blue HER at 50 mg/l concentrations of the dyes when used in a batch reactor (Shaffiq et al, 2002). Horseradish peroxidase has also been immobilized to be used for treating the effluents of paper and textile. Peralta-Zamora et al have recommended the photoenzymatic decolorization of textile effluents where the previous photochemical treatment of the effluent before exposure to the enzyme led to sufficient enhancement of the biological decolorization process (Peralta-Zamora et al, 1998). Gel entrapment methods have been tried out for immobilization of horseradish peroxidase and gel immobilized enzyme was found to be more efficient in dye removal than the free enzyme (Maddhinni et al, 2006). Reactions with enzymes have to be optimized with respect to pH, temperature, enzyme concentration and dye concentration values. The results with the removal of the dye Direct Yellow-12 in the presence of horseradish peroxidase indicated that the decolorization values obtained with the enzyme were lower above the dye concentration of 25 mg/l which can be acknowledged as the cut-off value. The optimum pH for dye removal by this enzyme was found to be 4.0 (Maddhinni et al, 2006). Matto and Hussain purified bitter gourd peroxidase by ammonium sulfate fractionation and further immobilized it on the surface of concavalin A layered calcium alginate-starch beads where it was found that the immobilized enzyme retained 69% of its original activity. Jack bean extract was used as a source of concavalin A which acts as a cheap source and helps in reducing the overall cost of the system. The authors have demonstrated very interesting studies on the effects of redox mediators on the effluent decolorization. The enzyme oxidizes the mediator and the oxidized mediator further oxidizes the substrate. Among the various redox mediators used, 1-hydroxybenzotriazole was found to be the most efficient, in the presence of which the free enzyme showed 28% decolorization and the immobilized enzyme showed 70% decolorization within 1 h. The authors further have also used a two reactor system for the decolorization of textile effluent, one reactor containing the immobilized enzyme and the other containing activated silica. This system was found to decolorize more than 90% of the textile effluent within 3 h of incubation in a batch process whereas the free form decolorized only 48% which might be because of the stability provided by immobilization techniques. Though the dye removal efficiency of the system was found to decrease with time even after 60 days the removal efficiency was 40% (Matto et al, 2009).

Salt fractionated bitter gourd peroxidase was also used for the decolorization of two water insoluble disperse dyes which were Disperse Red 171 and Disperse Brown 1. Here too, 1-hydroxybenzotriazole was found to be the most efficient redox mediator which facilitated 90% removal of Disperse Red 171 and 65% removal of Disperse Brown 1. Similarly, partially purified potato and brinjal polyphenol oxidases (PPO) have also been reported for the removal of textile and non textile dyes. Most efficient degradation with potato PPO was obtained at pH 3.0, while as the pH increased, the percentage decolorization values were found to decrease. Brinjal PPO showed no decolorization at pH 5.0. Thus, potato PPO mediated decolorization at broader pH ranges. The decolorization of dyes decreased after the time span of 1 h which may be because of the inhibition caused by the products. Potato PPO gave the highest decolorization (93%) for Reactive Blue 160 within 30 min of treatment with the enzyme. The enzymes were capable of decolorizing different dye mixture combinations containing 4 different dyes. A comparison of the two enzymes reveals that potato PPO's are more efficient than brinjal PPO's for the decolorization of textile dyes (Khan and Husain, 2007). Since purified enzymes can prove to be efficient systems for bioremediation, more work can be done in terms of the use of various inducers and redox mediators for making the decolorization processes more efficient. Moreover, if redox mediators are needed for better efficiency of the enzyme, an attempt could be made to use natural sources such as agricultural wastes as redox mediators. Similarly, cheaper processes and matrices should be tried out for immobilization which will help to reduce the overall cost of the dye removal process. Though purified and immobilized enzymes have shown the potential to degrade a variety of dyes, their use for phytoremediation is obviously costlier than the application of whole plant systems at the dye contaminated sites.

Table 1. The use of purified enzymes for the decolorization of dyes

Plant source	Enzyme	Purification techniques	Dyes decolorized	Reference
Ipomoea palmata	Peroxidase	Ion exchange chromatography, gel filtration chromatography	Methyl Orange, Chrysoidine, Supranol Green, Brilliant Green, Direct Blue, Crystal Violet	Shaffiq et al, 2002
Saccharum spontaneum	Peroxidase	Ion exchange chromatography, gel filtration chromatography	Chrysoidine, Blue MR, Porcion, Brilliant Blue HER, Supranol Green, Porcion Green HE-4BD, Direct Blue.	Shaffiq et al, 2002
Horseradish	Peroxidase	Dialysis	Direct Yellow-12, Direct Yellow 11, Basazol 46L, Azure B, Poly R-478, Remazol Brilliant Blue R, Crystal violet, Textile effluent.	Maddhinni et al, 2006; Knutson et al, 2005; Gramms and Rudeschko, 1998.
Soybean	Peroxidase		Direct Yellow 11, Basazol 46L, Azure B, Poly R-478, Remazol Brilliant Blue R, Crystal Violet.	Knutson et al, 2005; Gramms and Rudeschko, 1998.
Brinjal	Polyphenol oxidase	Ammonium sulfate precipitation, dialysis	Reactive Blue 160, Reactive Blue 171, Reactive Red 11, Reactive Orange 4, Reactive Yellow 84, Reactive Orange 86, Reactive Blue 4, Reactive Red 120, PAGE Blue 83, Commasie Brilliant Blue G 250, Comassie Brilliant Blue R 250, Methylene Blue, Naphthaquinone-4-sulphonic acid, Tropaeolin, Evans Blue, Dye mixtures	Khan and Husain, 2007

Table 1. (Continued).

Plant source	Enzyme	Purification techniques	Dyes decolorized	Reference
Potato	Polyphenol oxidase	Ammonium sulfate precipitation, dialysis	Reactive Blue 160, Reactive Blue 171, Reactive Red 11, Reactive Orange 4, Reactive Yellow 84, Reactive Orange 86, Reactive Blue 4, Reactive Red 120, PAGE Blue 83, Commasie Brilliant Blue G 250, Comassie Brilliant Blue R 250, Methylene Blue, Naphthaquinone-4-sulphonic acid, Tropaeolin, Evans Blue, Dye mixtures.	Khan and Husain, 2007
Bitter gourd	Peroxidase	Ammonium sulfate fractionation	Disperse Red 17, Disperse Brown 1.	Matto and Husain, 2009
Turnip (*Brassica rapa*)	Peroxidase	Ammonium sulfate fractionation	Acid Blue 92, Acid Red 97, Acid Yellow 42, Acid Black 1, Acid Black 210.	Kulshrestha and Husain, 2007.

Chapter 10

FUTURE PROSPECTS

The entire literature survey of dye decolorizing processes using plants makes one realize the immense potential of plants as systems for the efficient removal of dyes. Though plant systems have many advantages over bacteria that have been studied for dye degradation, there are many dimensions and processes involving the removal of dyes mediated by plants that are still unexplored. Hence, this area demands a lot of attention from researchers over the world for broadening the horizons of this technology.

Similar to metal phytoremediation, there is a need for extensive screening of different plant species that have the capacity to remove dyes that will help the application of more and more efficient systems for remediation of dyes. But, plant mediated removal of textile dyes has some limitations too. Unlike microbial degradation processes plant systems are known to be slower. Moreover, their capacity to tolerate high dye concentrations is also limited in comparison to microbes. Phytoremediation is restricted to the sites of contamination as deep as plant roots. Some of these limitations can be overcome by using genetic engineering techniques. Genetic modification techniques can be used to over express the enzymes involved in the existing metabolic pathways of the plant. Moreover, newer pathways can also be incorporated into the plants. If the plant does not have the ability to produce one or more enzymes that can degrade dyes or if the activities of these enzymes are low, related microbial genes can be inserted into the plant so as to make it a more potent system for phytoremediation. Transgenic poplar tress expressing the mammalian cytochrome P450 enzyme has been shown to be 640 fold faster than wild poplar trees for the removal of trichloroethylene. These enzymes also have a role in dye degradation.

Hence, such transgenic plants should also be tested for the removal of textile wastes (Aken 2008). Such genetic modifications can aim at achieving the complete mineralization of the dye molecules and probably help to make the plants tolerant to very high dye concentrations. In addition, similar to the studies involving plant-microbe associations for the degradation of textile dyes, we also need to consider approaches involving the synergistic and/or cumulative effects caused because of the use of different plant species (Cheema et al, 2010). Out of the many plants used for the degradation of textile dyes, no two plants can be found to be exactly similar with respect to characteristics such as enzymatic response, capacity for accumulation and/or adsorption of dyes etc. Combinatorial approaches using different plant species will therefore involve the merits of all the plant species used and can lead to enhanced degradation of the dye contaminated sites. Molecular studies need to be done to find out whether the induction in the enzyme activities upon exposure to dyes could be attributed to the over expression of the genes responsible for the production of these enzymes.

Thus, a more detailed understanding of plant processes for dye removal can help us to manipulate these systems in order to achieve more efficient remediation of the dye contaminated sites. Till today, many biological alternatives have been suggested for the removal of dyes from industrial effluents but due to various hindrances very few have actually been used on a large scale for the detoxification of effluents. Phytoremediation is a more applicable technology as compared to other bioremediation processes for dye removal and thus adequate research in this area will definitely lead to the actual application of these systems to remediate dye contaminated sites.

REFERENCES

Abadulla, E., Tzanov, T., Costa, S., Robra, K.H., Cavaco-Paulo, A., Gubitz, G.M., 2000. Decolorization and detoxification of textile dyes with a laccase from *Trametes hirsuta*. *Appl. Environ. Microbiol.* 66, 3357-3362.

Aken, B.V., 2008. Transgenic plants for phytoremediation: helping nature to clean up environmental pollution. *Trends Biotechnol.* 26, 25-27.

Aubert, S., Schwitzguébel, J.P., 2002. Capillary electrophoretic separation of sulphonated anthraquinones in a variety of matrices. *Chromatographia.* 56, 693-697.

Aubert, S., Schwitzguebel, J.P., 2004. Screening of plant species for the phytotreatment of wastewater containing sulfonated anthraquinones. *Water Res.* 38, 3569-3575.

Bafana, A., Jain, M., Agrawal, G., Chakrabarti, T., 2009. Bacterial reduction in genotoxicity of Direct Red 28 dye. *Chemosphere.* 74, 1404-1406.

Barbera, A.C., Cirelli, G.L., Cavallaro, V., Silvestro, I.D., Pacifici, P., Castiglione, V., Toscano, A., Milani, M., 2009. Growth and biomass production of different plant species in two different constructed wetland systems in Sicily. *Desalination.* 246, 129-136.

Bestani, B., Benderdouche, N., Benstaali, B., Belhakem, M., Addou, A., 2008. Methylene Blue and iodine adsorption onto an activated desert plant. *Bioresour. Technol.* 99, 8441-8444.

Carias, C.C., Novais, J.M., Martins-Dias, S., 2007. *Phragmites australis* peroxidases role in the degradation of an azo dye. *Water Sci. Technol.* 56, 263-269.

Carias, C.C., Novais, J.M., Martins-Dias, S., 2008. Are *Phragmites australis* enzymes involved in the degradation of the textile azo dye Acid Orange 7? *Bioresour. Technol.* 99, 243-251.

Chaudhry, Q., Zandstra, M.B., Gupta, S., Joner, E.J., 2005. Utilizing the synergy between plants and rhizosphere organisms to enhance breakdown of organic pollutants in the environment. *Environ. Sci. Pollut. Res.* 12, 34-48.

Cheema, S.A., Khan, M.I., Shen, C., Tang, X., Farooq, M., Chen, L., Zhang, C., Chen,Y., 2010. Degradation of phenanthrene and pyrene in spiked soils by single and combined plants cultivation. *J. Hazard. Mater.* 177, 384-389.

Chen, C.H., Chang, C.F., Liu, S.M., 2010. Partial degradation mechanisms of Malachite Green and Methyl Violet B by *Shewanella decolorationis* NTOU1 under anaerobic conditions. *J. Hazard. Mater.* 177, 281-289.

Chen, L., Flurkey, W.H., 2002. Effect of protease inhibitors on the extraction of crimini mushroom tyrosinase isoforms. *Curr. Top. Phytochem.* 5, 109-120.

Cluis, C., 2004. Junk-greedy Greens: phytoremediation as a new option for soil decontamination. *Bio. Teach. J.* 2, 61-67.

Cunningham, S.D., Ow, D.W., 1996. Promises and prospects of phytoremediation. *Plant Physiol.* 110, 715-719.

Davies, L.C., Cabrita, G.J.M., Ferreira, R.A., Carias, C.C., Novais, J.M., Martins-Dias, S., 2009. Integrated study of the role of *Phragmites australis* in azo-dye treatment in a constructed wetland: From pilot to molecular scale. *Ecol. Eng.* 35, 961-970.

Davies, L.C., Carias, C.C., Novais, J.M., Martins-Dias, S., 2005. Phytoremediation of textile effluents containing azo dye by using *Phragmites australis* in a vertical flow intermittent feeding constructed wetland. *Ecol. Eng.* 25, 594-605.

Doran, P.M., 2009. Application of plant tissue cultures in phytoremediation research: Incentives and limitations. *Biotechnol. Bioeng.* 103, 60-76.

Duc, R., Vanek, T., Soudek, P., Schwitzguebel, J.P., 1999. Accumulation and transformation of sulfonated aromatic compounds by Rhubarb cells (*Rheum palmatum*). *Int. J. Phytorem.* 1, 255-271.

EPA, 2000. Introduction to phytoremediation. EPA/600/R-99/107. National Risk Management Research Laboratory, Cincinnati, OH.

Ghodake, G.S., Telke, A.A., Jadhav, J.P., Govindwar, S.P., 2009. Potential of *Brassica juncea* in order to treat textile effluent contaminated sites. *Int. J. Phytorem.* 11, 297-312.

Gramss, G., Rudeschko, O., 1998. Activities of oxidoreductase enzymes in tissue extracts and sterile root exudates of three crop plants, and some properties of the peroxidase component. *New Phytol.* 138, 401-409.

Kagalkar, A.N., Jagatap, U.B., Jadhav, J.P, Bapat, V.A., Govindwar, S.P., 2009. Biotechnological strategies for phytoremediation of the sulphonated azo dye Direct Red 5B using *Blumea malcolmii Hook. Bioresour. Technol.* 100, 4104-4110.

Kagalkar, A.N., Jagatap, U.B., Jadhav, J.P., Govindwar, S.P., Bapat, V.A., 2010. Studies on phytoremediation potentiality of *Typhonium flagelliforme* for the degradation of Brilliant Blue R. Planta. 232, 271-285.

Kalme, S.D., Parshetti, G.K., Jadhav, S.U., Govindwar, S.P., 2006. Biodegradation of benzidine-based dyes Direct Blue 6 by *Pseudomonas desmolyticum* NCIM 2112. *Bioresour. Technol.* 98, 1405-1410.

Khan, A.A., Husain, Q., 2007. Potential of plant polyphenol oxidases in the decolorization and removal of textile and non-textile dyes. *J. Environ. Sci.* 19, 396-402.

Knutson, K., Kirzan, S., Ragauskas, A., 2005. Enzymatic bioleaching of two recalcitrant paper dye with horseradish and soybean peroxidase. *Biotechnol. Lett.* 27, 753-758.

Maddhinni, V.L., Vurimindi, B.H., Yerramilli, A., 2006. Degradation of azo dye with horseradish peroxidase (HRP). *J. Indian Inst. Sci.* 86, 507-514.

Matto, M., Husain, Q., 2009. Decolorization of textile effluent by bitter gourd peroxidase immobilized on concanavalin A layered calcium alginate-starch beads. *J. Hazard. Mater.* 164, 1540-1546.

Ncibi, M. C., Mahjoub, B., Seffen, M., 2007. Adsorptive removal of textile reactive dye using *Posidonia oceanica* (L.) fibrous biomass. *Int. J. Environ. Sci. Tech.* 4, 433-440.

Nilratnisakorn, S., Thiravetyan, P., Nakbanpote, W., 2007. Synthetic reactive dye wastewater treatment by narrow-leaved cattails (*Typha angustifolia* Linn.): Effects of dye, salinity and metals. *Sci. Total Environ.* 384, 67-76.

Ong, S.A., Toorisaka, E., Hirata, M., Hano, T., 2005. Treatment of azo dye Orange II in aerobic and anaerobic-SBR systems. *Process Biochem.* 40, 2907-2914.

Page, V., Schwitzguébel, J.P., 2009. The role of cytochromes P450 and peroxidases in the detoxification of sulphonated anthraquinones by rhubarb and common sorrel plants cultivated under hydroponic conditions. *Environ. Sci. Pollut. Res.* 16, 805-816.

Patil, P., Desai, N.S, Govindwar, S., Jadhav, J.P., Bapat, V., 2009. Degradation analysis of Reactive Red 198 by hairy roots of *Tagetes patula* L. (Marigold). *Planta* 230, 725-735.

Peng, S., Penga, S., Zhoua,Q., Caia, Z., Zhanga, Z., 2009. Phytoremediation of petroleum contaminated soils by *Mirabilis jalapa* L. in a greenhouse plot experiment. *J. Hazard. Mater.* 168, 1490-1496.

Peralta-Zamora, P., Esposito, E., Pelegrini, R., Groto, R., Reyes, J., Duran, N., 1998. Effluent treatment of pulp and paper and textile industries using immobilized horse radish peroxidase. *Environ. Technol.* 19, 55-63.

Rudrappa, T., Neelwarne, B., Kumar, V., Lakshmanan, V., Venkataramareddy, S.R., Aswathanarayaana, R.G., 2005. Peroxidase production from hairy root cultures of red beet (*Beta vulgaris). Electron. J. Biotechnol.* 8, 185-196.

Sarkanen, S., Razals, R.A., Piccariellos, T., Yamamotoa, E., Lewiss, N.G., 1991. Lignin peroxidase: toward a clarification of its ole *in vivo. J. Biol. Chem.* 266, 3636-3643.

Schwitzguébel, J.P., Aubert, S., Grosse, W., Laturnus, F., 2002. Sulphonated aromatic pollutants: limits of microbial degradability and potential of phytoremediation. *Environ. Sci. Pollut. Res.* 9, 62-72.

Senan, R.C., Abraham, E.T., 2004. Bioremediation of textile azo dyes by aerobic bacterial consortium. *Biodegradation.* 15, 275-280.

Shaffiqu, T.S., Roy, J.J., Nair, R.A., Abraham, T.E., 2002. Degradation of textile dyes mediated by plant peroxidases. *Appl. Biochem. Biotechnol.* 102-103, 315-326.

Sharma, K.P., Sharma, K., Kumar, S., Grover, R., Soni, P., Bharadwaj, S.M., Chaturvedi, R.K., Sharma, S., 2005. Response of selected aquatic macrophytes towards textile dye wastewaters. *Indian J. Biotechnol.* 4, 538-545.

Sharma, P., Goel, E.R., Capalash, N., 2007. Bacterial laccases. *World J. Microbiol. Biotechnol.* 23, 823-832.

Strycharz, S., Shetty, K., 2002a. Peroxidase activity and phenolic content in elite clonal lines of *Mentha pulegium* in response to polymeric dye R-478 and *Agrobacterium rhizogenes. Process Biochem.* 37, 805-812.

Strycharz, S., Shetty, K., 2002b. Response of oregano (*Origanum vulgare* L.) clonal lines to *Pseudomonas* sp. Z strain and polydye R-478 and implications for hyperhydricity prevention in tissue culture. *Process Biochem.* 38, 343-350.

Takahashi, M., Tsukamoto, S., Kawaguchi, A., Sakamoto, A., Morikawa, H., 2005. Phytoremediators from abandoned rice field. *Plant Biotechnol.* 22, 167-170.

Togo, C.A., Mutambanengwe, C.C.Z., Whiteley, C.G., 2008. Decolorisation and degradation of textile dyes using a sulphate reducing bacteria (SRB)-biodigester microflora co-culture. *Afr. J. Biotechnol.* 7, 114-121.

Vangronsveld, J., Herzig, R., Weyens, N., Boulet, J., Adriaensen, K., Ruttens, A., Thewys, T., Vassilev, A., Meers, E., Nehnevajova, E., Lelie, D., Mench, M., 2009. Phytoremediation of contaminated soils and groundwater: lessons from the field. Environ. *Sci. Pollut. Res.* 16, 769-794.

Vidali M., 2001. Bioremediation: An overview. *Pure Appl. Chem.* 73, 1163-1172.

Weyens, N., Taghavi, S., Barac, T., Lelie, D., Boulet, J., Artois, T., Carleer, R., Vangronsveld, J., 2009a. Bacteria associated with oak and ash on a TCE-contaminated site: characterization of isolates with potential to avoid evapotranspiration of TCE. *Environ. Sci. Pollut. Res.* 16, 830-843.

Weyens, N., Lelie, D.,Taghavi, S., Newman, L., Vangronsveld, J., 2009b. Exploiting plant-microbe partnerships to improve biomass production and remediation. *Cell.* 27, 591-598.

Zabłudowska, E., Kowalska, J., Jedynak, L., Wojas, S., Skłodowska, A., Antosiewicz, D.M., 2009. Search for a plant for phytoremediation-What can we learn from field and hydroponic studies? *Chemosphere.* 77, 301-307.

INDEX

A

B

C

D

E

F

G

H

I

K

L

M

N

O

P

Q

R

S

T

U

V

W

X

Z